STEAM WAGONS IN PRESERVATION

MALCOLM BATTEN

First published 2024

Amberley Publishing
The Hill, Stroud
Gloucestershire, GL5 4EP

www.amberley-books.com

Copyright © Malcolm Batten, 2024

The right of Malcolm Batten to be identified as the Author of this work has been asserted in accordance with the Copyrights, Designs and Patents Act 1988.

ISBN 978 1 3981 2016 7 (print)
ISBN 978 1 3981 2017 4 (ebook)

All rights reserved. No part of this book may be reprinted or reproduced or utilised in any form or by any electronic, mechanical or other means, now known or hereafter invented, including photocopying and recording, or in any information storage or retrieval system, without the permission in writing from the Publishers.

British Library Cataloguing in Publication Data.
A catalogue record for this book is available from the British Library.

Typesetting by Hurix Digital, India.
Printed in the UK.

Contents

	Introduction	4
1	Flashback – Working Days	7
2	Replicas of Early Designs	8
3	Overtype Wagons	9
4	Undertype Wagons	46
5	Transverse Boiler Wagons	81
6	Home-made Freelance Wagons	84
7	Miniatures	91
	Acknowledgements, Bibliography and Further Reading	96

Introduction

The steam lorry (or wagon as it is more commonly known) dates back to the origins of steam motive power. French engineer Nicholas Cugnot first proposed a steam-powered 'Fardier à Vapeur' road vehicle for artillery haulage in 1763. With patronage from the Minister of War and public funds, he initially demonstrated a machine in October 1769; it achieved a speed of 2 mph, but the boiler could only run for around fifteen to twenty minutes before having to pause to build up steam pressure. A second machine of 1769 with an improved boiler feed pump was demonstrated in Paris in June 1770, and possibly a third machine in 1770–71. Unfortunately, after the successful demonstration and while this vehicle was being towed home by horses it ran out of control down a hill and crashed into the arsenal wall. The project was abandoned but parts of the vehicle did survive and these were matched with parts from the earlier 1769 version and preserved at the Conservatoire national des arts et métiers (Cnam) museum in Paris. Cugnot's wagons had three wheels, with the steerable front wheel driven by two pistons connected to a copper boiler mounted ahead of this wheel. A working replica has been constructed using the original working notes and drawings and tools available at Cugnot's time.

In Britain, the first successful attempt at a steam road vehicle was William Murdock's three-wheel steam carriage of 1786. But the main breakthrough was by the celebrated Cornish engineer Richard Trevithick, who pioneered the application of high-pressure steam. After building some models, he built the full-size *Puffing Devil* in 1801, which successfully climbed a hill in Camborne on 28 December. In 1802 Trevithick obtained a patent which described 'Methods of improving the construction of steam engines and the application thereof for driving carriages and for other purposes'. A working replica of Trevithick's 1801 *Puffing Devil* was built in 2001 by the Trevithick Society for the bicentenary. This has been demonstrated at various locations including in France in 2013 where it was displayed alongside the replica Cugnot wagon.

From the 1820s to the 1840s there were several attempts at designing steam carriages, the most successful of which were those by William Henry James, Sir Goldsworthy Gurney and Walter Hancock. But development was hampered by hostile legislation and also by the poor state of Britain's roads at the time.

The traction engine was developed in the mid-nineteenth century and had evolved into an established form by the 1870s. Road locomotives became a specialised form of traction engine, with road springs, extra belly water tanks, headlights and a canopy for the crew. These were designed for heavy road haulage of trailers for loads that might be too large for carriage by rail. These engines often worked in pairs on the front and another behind to provide braking on gradients. In the early 1900s smaller versions known as tractors were developed for hauling lighter trailer loads such as furniture vans or tree trunks (round timber).

The steam wagon was also developed from the 1890s but now had another rival to the horse as the internal combustion engine had also been invented. The early years of the twentieth century saw both forms being improved. Various trials were conducted by the military authorities in the run-up to the First World War, which was the first mechanised war and saw widespread use of both steam- and petrol-driven vehicles.

Before the early 1920s the legal speed limit for wagons on solid rubber tyres was 12 mph or 5 mph with a trailer, although this was usually exceeded. The 1920s saw the Depression and some manufacturers went to the wall as demand declined.

In the 1930s the more efficient diesel engine began to take over from petrol for the larger trucks. Both steam wagons and petrol or diesel trucks were improved, and now pneumatic tyres became the norm rather than the former solid tyres. However, there could be only one winner, and by the mid-1930s with punitive axle weight road tax discriminating against steam, the end was in sight for steam wagons and traction engines generally. One company, W. J. King of Bishops Lydeard, Somerset, laid aside nine Foden tipper wagons in 1931, the oldest no more than six years old. These remained stored, many in the open, for the next fifty-seven years until the rusting remains were sold for preservation in May 1988. The entry into the British market of Ford and Bedford with their cheap mass-produced light trucks also spelled the end for several of the smaller British truck makers. The steam wagon had been a largely British product, encouraged by an abundance of cheap coal. It never caught on anywhere to the same extent it did in Europe or America.

While most of the main makers of traction engines also built steam wagons, for most this was a minor part of their production. There were, however, two main manufacturers of steam wagons: Foden, from Sandbach, and Sentinel of Shrewsbury. These had championed contrasting designs. Foden had produced the 'overtype' wagon with a horizontal front boiler and cylinders mounted above, as in the style of other traction engine types. Drive was by chain to the rear axle. This pattern was adopted by many of the other makers. Sentinel opted for the 'undertype' design. This had a vertical front-mounted boiler sited either ahead of or behind the driver, and the cylinders mounted underfloor. Drive was initially by chain, but later shaft-drive engines took over. This arrangement took up much less room, so allowed for more load space. Eventually even Foden turned to the undertype but by then the tide had turned and only 134 were built. One other alternative arrangement was adopted by the Yorkshire Patent Steam Wagon Co. Ltd with a transverse mounted horizontal front boiler.

Both Foden and Sentinel also built specialist timber tractors which were designed for forestry work and had a steam-powered winch for loading the logs onto trailers.

In 1932 the Foden company split when Edwin R. Foden set up on his own to make diesel lorries utilising Gardner engines. The main company had built their first diesel lorry in 1931 and ended steam production in 1934. Sentinel ceased home production in 1938 but gained an export order for 100 S6 tippers from Argentina in 1949. They began making diesel vehicles from 1947 and unusually they kept the underfloor engine position in these and had the distinction of being the first British manufacturer to offer an underfloor-engined bus chassis on the market. However, the company were bought out by Rolls-Royce in 1956 and thereafter the Shrewsbury site made diesel engines for them.

There were also steam buses. In 1909 Thomas Clarkson formed the National Steam Car Co. Ltd, which both made and operated steam buses in London. When production stopped in 1914 there were 173 of these paraffin-fired buses in use. In 1920 Clarkson also introduced a 3-ton coke-fired wagon design, which had an automatic feed-water system and a condenser. No complete Clarkson vehicle survives in the UK.

By 1960, the commercial life of steam wagons had almost come to an end. A few had a reprieve in the 1950s when the Suez crisis hit oil supplies. Others hung on for instance as tar sprayers where their boilers kept the tar warm. The last commercial user was Brown Bayley Steelworks, Sheffield, who had a number of Sentinel wagons in internal use until at least 1968. These were used to supply the open-hearth furnaces with scrap and take ingots to the rolling mill. They had survived in use because they were on solid rubber tyres which was more suitable for the harsh terrain with red hot metal fragments.

While most wagons went for scrap, hundreds have survived into preservation to tell their story. In the 1950s, engines could be obtained for little more than their scrap value – sometimes less than £25. While inevitably the vast majority of the survivors are Fodens and Sentinels, there are examples, sometimes just a single representative, from a range of the other makes. Some of these have survived by dint of being repatriated from export markets, such as the only example of an Atkinson wagon. There are, though, some makers, notably Allchin, from which there are no survivors, at least not in the United Kingdom. A visit to any traction engine rally is likely to produce some wagons, probably with both Foden and Sentinel vehicles in attendance.

Possibly the greatest ever gathering of steam wagons came at the 2012 Bedfordshire Steam & Country Fayre at Old Warden when nearly sixty wagons and timber tractors from twelve different manufacturers took part in a Steam Wagon Spectacular.

The Traction Engine Register is the authoritative listing of all known traction engines in the United Kingdom and Eire, including steam wagons, fire engines and portable engines. There are over 4,000 engines listed in total, and this is updated every four years.

Also included in this book are a selection of freelance steam wagons and vans that have been created, often using old motor vehicle chassis. Also, a selection of the many fine miniature versions of steam wagons that have been made and which can be regularly seen at rallies alongside their full-size cousins.

All photographers are by the author except where stated.

1
Flashback – Working Days

Many engines now in preservation were still working for their living in the early 1950s. Some steam wagons finished life as tar sprayers, the steam keeping the tar hot for pouring. This is Sentinel DG6 No. 8562, FD 6603 of 1931, which was with Thames Tar Products and Contractors Ltd at their Beddington Lane Depot near Croydon on 2 December 1950. (Photo by John Meredith/Online Transport Archive)

Another Sentinel, North Thames Gas Board No. 267 CGX 845, a later S4 shaft-driven type taken in Westminster Bridge Road on 9 November 1950. Unlike FD 6603, this has not survived. (Photo by John Meredith/Online Transport Archive)

2
Replicas of Early Designs

A working full-size replica of a Walter Hancock 1833 steam carriage design, *Enterprise*, was completed in 2003. Registered Q331 RMA, it is seen attending the London Transport Museum Depot at Acton in 2014. The engine is at the rear with the water tank mounted behind the rear wheels.

The front end of the steam carriage. The steersman is mounted somewhat precariously on a seat that is on a platform projected from the front of the body.

3
Overtype Wagons

Aveling & Porter (Aveling & Porter Ltd, Rochester, Kent)

Only one example of a wagon by Aveling & Porter survives in the United Kingdom and Eire, according to the Traction Engine Register. No. 9282 was built in 1922 and was first preserved as part of the Tom Varley collection. Here it is seen at the East of England Showground at Peterborough in August 1979.

D 3777 is seen again, in later ownership and lettering, at the Great Dorset Steam Fair in 1994. This is a Colonial type and was sent new to Australia for Gudgegory Council near Sydney, later passing to Mudgee Council. The wagon was repatriated in 1978.

Clayton & Shuttleworth (Clayton & Shuttleworth Ltd, Lincoln)

Clayton & Shuttleworth first made wagons in 1912. Over 300 of their wagons were used by the War Department in the First World War. No. 48347 (FE 2931) dates from 1919. This was part of the collection of Robert Crawford at Frithville, Boston. This was also seen at the East of England Showground, Peterborough, taken in 1980. Note that Aveling & Porter D 3777 is parked up behind it.

In 2000, Clayton & Shuttleworth wagon No. 48510 (FE 3344) *The Fenland Princess* waits to enter the ring at the Bedfordshire Steam & Country Fayre, Old Warden. This dates from 1920. It spent much of its working life with Tarroads as a tar sprayer, eventually ending up in Hardwicke's scrapyard, Ewell, in a derelict condition.

No. 48637 also dates from 1920. This worked and was later preserved in Ireland until 2010 where it was registered MI 946. It was a visitor to the Great Dorset Steam Fair (GDSF) in 1994 where it is shown here.

In 2010 the wagon changed hands and came back to England, where it regained its original registration FE 3704. This shows it attending the Netley Marsh Rally near Southampton in 2014. Note the somewhat precarious steering position. The unguarded flywheel is also clearly visible – no health and safety laws in those days!

Foden (Fodens Ltd, Sandbach, Cheshire)

Edwin Foden's first steam wagon was produced in 1900. By 1903 the design had standardised into a compound engine with two speeds and a single chain drive. Several thousand of these were made including some 800 for the British Army. In 1922 the design was replaced by the three-speed C type, which also featured a raised driving position for better visibility and better steering and braking. A three-axle model for 12-ton loads was also offered and an articulated model.

Oldest of the Foden wagons being rallied was No. 1364 of 1907. This was recorded at Blackpool in 1991. However, this wagon is listed as having been exported to Sweden in 1993. An older wagon, No. 848 of 1904, is on static display at the Thinktank Birmingham Science Museum.

Also from 1907 is No. 1427 *Isabella*, now registered BF 5148. This was repatriated from Chile. Taken at the rally at Preston Services, Kent, in 2010.

No. 2876 (M 3631), a 4-ton wagon from 1911, worked for haulage company Pickford's until the early 1920s. It was then laid aside in a yard at Gosport until rescued in 1980. This was how it looked in 1983 at Knowl Hill, near Maidenhead.

By 1993 M 3631 was looking somewhat different, seen here at Brighton at the end of the HCVS London to Brighton Historic Commercial Vehicle Run.

At the Netley Marsh Rally in 2019 Foden wagon M 4257 *Tewlass*, No. 3210 of 1912, was a work in progress. The overtype cylinder block and flywheel are clearly visible. This is recorded as being reconstructed from a tractor.

Foden wagon No. 3510 (M 4673) has an interesting story. Built in 1913 for J. T. Lawton Ltd of Manchester, it was purchased in 1928 by Henry Ford for his museum at Dearborn, Michigan. It would remain static there until repatriated and returned to steaming order in 1980. When seen in 1983 it still carried its original paintwork, its long period in the museum having prevented deterioration. The location was Hyde Park, London, during a so-called Victorian Parade.

Making a first public appearance after restoration at the Expo Steam show at Peterborough in 1980 was Foden Colonial No. 4086 of 1913, which had been repatriated derelict from Australia the previous year. The paint was barely dry! The boiler of this design was modified for wood burning. This wagon was later re-exported to Germany where it spent the years 1984 to 2008. This was from the extensive collection built up by the late Tom Varley.

No. 6368 *Pride of Edwin* of 1916 was retained by Fodens Ltd. After the company was sold to Paccar in 1980 it later passed to the Science Museum, where it is kept in the reserve collection at Wroughton Airfield near Swindon. It was photographed at a rare open day held there on 11 September 1988.

No. 7536 was new in 1917 to the War Department and then sold to Devon County Council. When photographed at the Royal Gunpowder Mills, Waltham Abbey, in 2007 it had recently been repainted in Devon CC colours.

From 2014 to 2018 the nation was celebrating the centenary of the First World War, and the Great Dorset Steam Fair took a major role by creating a trenches experience area. There were trenches, field workshops, artillery emplacements, etc., all enhanced by vehicles of the period. Foden wagon No. 7768 (M 8562) dating from 1917 is seen among a selection of contemporary lorries in 2017. This tipping wagon originally saw service in France during the First World War, being used on repairs to roads following shelling damage. After the war it was sold to a French company from Cambrai where it was used until 1948. The wagon is fitted with buffers either side of the smokebox for shunting railway wagons.

No. 10694 (PC 9928) *Tiler* was built as a 5-ton wagon in 1923 and supplied to W. Young & Son, Purley, Surrey. The name derives from its later working life with the Maidenhead Brick & Tile Co. This was taken at the Great Dorset Steam Fair in 1990.

No. 13388 (TU 3113) is a 6-ton wagon supplied in 1926 to W. J. King of Bishops Lydeard. This was one of the nine Foden tippers laid up in 1931 where they remained until auctioned, very derelict, in 1988. This was then the first of these to be fully restored. Seen at Tallington in 1997.

Fitted with a brewer's dray body is TP 9530, No. 13624 of 1930. This was seen at Horstead Keynes on the Bluebell Railway in 1994. At the time this was incorrectly believed to be No. 13042.

What is the connection between this wagon and the previous photograph? They are actually the same wagon. Now correctly identified as No. 13624 and registered WX 2682, this is believed to be the last chain-steered Foden made and was one of two built for the West Riding of Yorkshire as a three-way tipper. It is now fitted with a 1901 Liftvan container and painted in the colours of Bishop & Sons, who operated similar Foden wagons. It has been a regular participant on the London–Brighton Run where it was seen arriving in 2004.

No. 13708 was delivered new in 1930 to May, Gurney & Co. Ltd, Norfolk, and later adapted as a tar sprayer. Withdrawn for scrap in 1955, it then passed to George Cushing before moving on to Alan Bloom as part of the Bressingham Steam Museum collection in 1962. Last steamed there in 1972, it later passed to the Saunders collection at Stotfold where it returned to steam in 2012. In 2021 it revisited Bressingham as part of the sixtieth anniversary celebrations.

Taken at Clapham Common, London, in 1973 was 1931 Foden No. 13848 (RB 3525). New to Derbyshire County Council with a tipper body, it finished commercial life as a tar sprayer and first passed into preservation in 1955. When seen here it had a steel body from an ex-military Bedford lorry and was painted in Charringtons Transport livery. I would subsequently photograph this with two quite different bodies.

At Brockwell Park, London, in 1977 was No. 13848 again – at this time it was carrying this tar tanker body and in Chart Carrying Co. livery.

When photographed at Brighton in 1985 RB 3525 had been fitted with a lift van body lettered for Radcliffe & Son, carriers and storers. The Foden was now owned by Geoff Radcliffe.

RB 3525 again. This is now resident on the Isle of Man but was visiting a rally at Carrog on the Llangollen Railway when photographed in 2001.

Also fitted with a box van body is WL 910 *Britannia*, No. 12228 of 1926. As the lettering states this was fitted *c*. 1967 for a round-the-world journey in 1968–72 by owner Michael List-Brain. The only steam vehicle to achieve such a feat, countries visited included Australia, India and the United States. It is normally resident at Preston, Kent, but was visiting Chatham Dockyard on this occasion in 1990.

Dating from 1930, RP 9208 (No. 13716) is fitted with a three-way tipping body, being demonstrated in this view taken at Ardingly in July 1984. It was new to Northamptonshire County Council and was rescued from a scrapyard at Ely.

No. 11340 of 1923 was originally a dray wagon for a Brighton brewery registered CD 8223 and was later with travelling showmen from Southampton. In the 1960s it was fitted with a replica body to that once carried by a Foden Motor Works Brass Band bus and took the registration M 6359 and name *Puffing Billy* of the band bus. When seen at Tallington in 1995 it was with an owner from County Durham after an earlier period as a static exhibit at the Bygone Village, Fleggburgh, near Great Yarmouth.

After periods with owners from Norfolk and County Durham, M 6359 is now with the Searle family collection from Horsham and is a regular attender at events, such as here at Brighton.

Also fitted with a bus body to a freelance design is No. 4258 (M 5798) of 1914. This was fitted in 2001 and the wagon is seen at the Great Dorset Steam Fair in 2003. Originally it was a 4-ton tipper with Samuel Allsopp & Sons, Burton on Trent.

TW 4207, No. 12364 of 1926, is representative of later wagons fitted with pneumatic tyres. Note also how the cab area has now become enclosed to offer better protection to the crew. This is seen at Peterborough in 1981 when in the colours of H. E. Parkin & Son. The present livery promotes Newquay Steam Beers. This 6-ton wagon started out as a three-way tipper with E. B. Devenish of Rayleigh, Essex, and ended its commercial life in 1962 as a tar sprayer.

AN 9568, No. 13138 of 1928, was already a seasoned attendee of rallies when photographed at Brockwell Park, London, in 1986 – note the display of rally plaques. It attended the first Bedford Steam Rally at Woburn Abbey in August 1957. Named *London Pride* after the beer of the same name, this was in the colours of Fullers Brewery at Chiswick, its original owners, who ran it until 1948. It was then purchased for preservation by T. T. Boughton & Sons of Amersham but has changed hands since then.

C type 6-ton dropside wagon PW 3767 of 1924 (No. 11538) carries the livery of its original owner from Norfolk. It carried farm produce to the station and coal on the return journey until retired in 1945. This was in the USA from 1963 to 1992 but now lives in Sussex where it was seen in 1996.

There are wagons from three different makes in this photograph, but taking centre stage is Foden No. 13802 of 1931, registered VN 2912. Seen at the 2001 Weeting rally, this is a resident at the Strumpshaw Steam Museum in Norfolk.

At the end of the 1976 HCVS London to Brighton Run, onlookers watch as 1929 Foden steam wagon RO 6330 is prepared for winching onto a low loader trailer for its journey home. Originally bought by Hertfordshire flour millers, this wagon was later sold to Tarroads Ltd and converted to a tar spraying unit, in which form it lasted to the mid-1950s.

No. 13764 (TF 3106) dates from 1930. This was taken at Enfield in 1992 in the green livery it carried at the time. This was new to Riding & Gillow before being converted to a tar sprayer for The Mechanical Tar Spraying & Grouting Co. Ltd of Reading.

By 1999 TF 3106 was repainted in the colours of The Mechanical Tar Spraying & Grouting Co. Ltd. Cab windows have been fitted and the flywheel encased. Here it is at Stotfold Mill in 2012. This wagon is part of the extensive Saunders family collection.

A well-known wagon in the south-west, No.13316 of 1929 carries a tanker body fitted *c.* 1984. It had ended life as a tar tanker but in early preservation the original tank had been replaced by a ballast box. When seen here in 1988 at the Great Dorset Steam Fair it was liveried for Horrell's cider, but now resides near Southampton and is lettered for the current owner, who has fitted period-style window frames.

Foden D type timber tractor No. 12370 *Island Chief* of 1926 passes behind the ring and the terrace stand at Rushmoor Arena in 1987. This was new to Arthur Kirby of Newton Abbot sawmills. It later passed to showmen Harris & Sons of Ashington, West Sussex. It later spent some time in the Hollycombe Steam Collection and was then with Charlie Russell of Bordon.

No. 12370 again after changing owners twice more and being repainted in the original colours, taken at Knowl Hill in 2002. Note the flywheel brake.

The same engine but after being rebuilt with this more enclosed cab design and platform over the rear wheels, which is how it was originally supplied when new. The original-style signwriting has also been reproduced on the cab sides. Woodcote Rally, 2009.

No. 13068 (EU 3764) *Perseverence* dates from 1928. It had a long and varied working career, starting out on timber haulage with an owner from the Brecon area, later passing to Langley & Johnson of Slough. During the Second World War it was employed on the demolition of bombed buildings in London. This view taken at Little Wymondley, Herts, in 1990.

Seen at Netley Marsh in 2014 sporting a recent repaint to the original owner's livery is 1929 Foden timber tractor No. 13444 *Little Lady*. From 1968 to 2014 this had carried showman's fittings.

Running through the back streets of Birmingham during the 1990 road run at the Birmingham Museum of Science & Industry (now closed) is No. 14084, believed to be the last Foden steam vehicle for the home market. Despite dating from 1933, AMB 300 looks much older due to it still sporting solid tyres rather than pneumatics. It was supplied new to J. Murch, Umberleigh, Devon, and had two further owners before being laid aside in 1948.

Now equipped with pneumatic tyres on the front and solids on the back is DF 4055 *Angelina*, No. 12782 of 1927. It originally had solid tyres all round and was a timber tractor with A. W. Brunsdon, Stonehouse, Gloucestershire. This has also been fitted with showman's twisted brass fittings and a dynamo – these being added during preservation days. It is seen in the ring at the 1995 Shrewsbury Rally.

UR 1328 *Merlin* (No. 13156) is not all it seems. Built in 1928 as a six-wheeled wagon, the remains of this were made up into this form *c.* 1990 and fitted with a new boiler. Old Warden, 2007.

The lettering says it all. LG 8784 is a Foden C type estate tractor. Although listed as No. 13484 of 1930, the Traction Engine Register states that it was built in 1926 as No. 12300 and carries this number on its works plate. It was built as an experimental half-track but rebuilt to a tractor in 1930 so perhaps Foden's allocated a new maker's number at rebuilding. Taken at the Royal Gunpowder Mills, Waltham Abbey, in 2008.

Now mounted on pneumatic tyres, tractor BUP 71 *Cestria* (No. 13218) dates from 1929. This was originally built as a six-wheel flat wagon and allocated registration ALG 274 but the intended sale fell though. It languished at Foden's until sold in 1936 to James Dixon of Wolsingham, County Durham, who had it shortened to a tractor with winches for hauling timber and it gained the present registration. The Foden, by then derelict, was bought for preservation by Henry Thompson in 1954 and restored three years later. He decided that it was difficult roading it to rallies in this form and shortened it to a four-wheel tractor on pneumatic tyres. This is seen at the appropriate local location of Beamish Museum with a backdrop of the coal mine pithead gear.

Elaborately lettered and lined out is VA 2519 *Hielan Laddie*, No. 11444 of 1924. Hailing from Aberdeen, it had travelled south to Chester-le-Street, County Durham, to attend a rally there in 1993. This started as a 6-ton wagon with Lanarkshire County Council.

KX 3340 *Samantha* is probably one of the most widely rallied wagons on the rally circuit. Here it is with a timber trailer at the Great Dorset Steam Fair in 1995. Although now in tractor form, it was supplied in 1929 as a tipping wagon to Buckinghamshire CC. It was later sold to T. T. Boughton & Sons of Amersham where it was rebuilt as a tractor and fitted with a winch for timber work in 1933.

In the ring at the Chiltern Steam Rally in 2014 we see Foden products of both steam and diesel types. AMB 300 the timber tractor from 1933 had now been repainted in its original owner's livery. Alongside is GV 8092, a 1941 Foden STG5 timber tractor which is fitted with a Gardner 5LW engine. The Chiltern Steam Rally at Prestwood near High Wycombe is in an area famous for timber and chair-making and usually produces a fine display of timber-handling vehicles and machinery.

Foden articulated tractor No. 13536 was entered in the Bedfordshire Steam & Country Fayre in 2016. This had been rebuilt to its original form in 2015 after many years in preservation with a previous owner as a tractor. Built in 1930, the original commercial owner was F. H. Richards of Leicester.

No. 3398 (M 4489) of 1912 is listed as a 'wagon-rebuilt' in the Traction Engine Register 2020 edition, which says it has a tilt cover van bodywork. However, in this view from 2007 it appears to be a ballast box tractor. This was taken at Kemble in 2007.

In the collection of the late Robert Crawford, Frithville, Boston, was this rare Foden Sun type tractor. No. 13730 dates from 1930 and is one of only two of this type preserved in the UK and Ireland. It was photographed at the Peterborough Expo Steam held at the East of England Showground in 1979. The Traction Engine Register reports that it was rebuilt as a C type wagon by 1995.

Foden No. 13008 was built in 1928 as a rigid six-wheel wagon registered TU 9235. It was later converted to a tar sprayer and as such worked for the Limmer & Trinidad Lake Asphalt Company. It was converted to the present shorter tractor form in the early 1960s. This came to the Saunders collection in 1987 and is seen here at Great Amwell, Hertfordshire, in 1993 in the then blue livery.

At Old Warden in 2014 Foden No. 13008 displays its revised livery style as when it worked for The Limmer & Trinidad Lake Asphalt Company. Now registered SS 9191, it does not usually display this.

Foster (William Foster & Co. Ltd, Lincoln)

This is 1920 Foster 5-ton wagon No. 14550 carrying names *Sir William* and *Tritton*. This was brought back from Australia in 1983 by the late Tom Varley. Foster & Co. built just sixty steam wagons and this is the only survivor. Taken at the Chiltern Steam Rally, Prestwood, in 2003.

The same engine a year later following a repaint and bodywork modifications. Taken at the Bedfordshire Steam & Country Fayre, Old Warden, in 2004.

Garrett (Richard Garrett & Sons Ltd, Leiston, Suffolk)

No. 30826 (BJ 1366) was built in 1912 and sold to Charles Marston, a miller. It later to passed to the Gray family of Ipswich who converted it into a showman's wagon with a front dynamo. In 1926 it was in the ownership of Taylor Bros, Wimbish, Essex, who used it as a mobile welding vehicle. It was partially scrapped in the 1930s, but the chassis and rear axle assembly survived as part of a shed. The parts were acquired for restoration in 1975, changed hands in 1998, and purchased by the Worbey family in 2002. Restoration then started with the work being completed in 2011. This is the only 3-ton example preserved. Stotfold Mill, 2012.

Garrett wagon SX 2395 was in Canada until 2012 and is the only 6-ton overtype superheated steam wagon left. This was seen at the 2017 Great Dorset Steam Fair.

Hindley (E. S. Hindley & Son, Bourton, Dorset)

Some of the early manufacturers soon lost out to their more successful competitors. One such was E. S. Hindley & Son of Bourton, Dorset. FX 33 is a replica of one of their early wagons completed in 2010 and using some original parts, including the wheels. This was on show at Old Warden in 2012. There is no protection for the driver – they had to be pretty hardy in those days!

Mann (Mann's Patent Steam Cart & Wagon Co. Ltd, Hunslet, Leeds)

Mann's Patent Steam Cart & Wagon Co. Ltd from Hunslet, Leeds, specialised in wagons and tractors, and most of their surviving products are such. The oldest survivor is No. 881 of 1914. Seen at Preston Services in 2011, this was originally built as a tractor, later rebuilt as a roller and then to the wagon form seen here. This was later exported to the Czech Republic by 2015.

No. 1120 of 1916 carries a furniture van body and the livery of its original owners. This was rescued in 1960 and is seen at Old Warden, 2008.

At Prestwood in 2005 is rare Mann wagon U 4990 of 1919. The Mann engines were a product of Leeds, and this wagon like most carries a Leeds registration number. Mann wagons were produced from 1913 until 1929. This example was new to Beyer Peacock, Manchester. In 1950 it was converted to a steam sterilising plant and restoration required much sourcing and manufacture of parts.

U 4943 is an example of the Mann steam cart dating from 1919. Note the very narrow track of the front wheels. Seen at the Great Dorset Steam Fair in 1994.

Another steam cart, but U 4298 of 1917 looks more stable and substantial than the later U 4943. This was an entrant at Rushmoor Arena in 1994 and is based at the Hollycombe Steam in the Country Museum at Liphook. These engines were designed for one-man operation and could be used for direct ploughing (i.e. towing a plough), as well as carrying goods in the cart body.

Ransomes, Sims & Jefferies (Ransomes, Sims & Jefferies Ltd, Ipswich, Suffolk)

Ransomes, Sims & Jefferies wagon No. 34270 was built in 1923 and exported to Australia as a tipper. The remaining parts were repatriated in 1999 and the rebuild has included parts from a chassis that came from New Zealand and the construction of new running gear to create the only wagon by this make in the Traction Engine Register. I first photographed the completed wagon at Prestwood in July 2003.

When photographed at Preston, Kent, in 2008 the body had been rebuilt as a dray complete with beer barrels load.

Robey (Robey & Co. Ltd, Lincoln)

The theme for The Great Dorset Steam Fair in 2019 was engines made by the Lincoln-based companies of Clayton & Shuttleworth, Foster, Ruston Proctor, Ruston & Hornsby and Robey. Inside the Robey display marquee is wagon No. 42657 of 1925. This was new to Highways Colloidal Ltd and has been with Robert Crawford of Boston since 1966.

The only other wagon in the UK is No. 42522, which is fitted with a stayless circular 'thimble' firebox. It was new to William Gossage, Widnes, as an articulated tanker. Owned by the Robey Trust, it was rebuilt with parts from No. 42287, needing a new chassis, body, cab and steering gear. Restoration was completed in 1996. This appeared at the Great Gathering of wagons at Old Warden in 2012.

No. 43388 is an example of the Express tractor type of which only nine were built and two survive. VL 983 was built in 1929 and had four successive commercial owners. This was an exhibit at the 2003 Great Dorset Steam Fair. It had previously been with Tom Varley who named it *Pendle Knight*.

Tasker (W. Tasker & Sons Ltd, Andover, Hampshire)

The Hampshire County Museum Service display at the 1991 Rushmoor Arena show featured the sole surviving Tasker wagon No. 1915, making its debut after a six-year restoration. Built in 1924, this worked for W. J. King, quarry owners of Somerset. It was rescued from their quarry near Minehead in 1957. Behind are Tasker stationary engine No.111 of 1872 and portable engine No. 1228 of 1898. All these were saved by the Tasker Trust and maintained by the Hampshire County Museum. Since 2000 they have been kept at the Milestones Museum in Basingstoke.

The Tasker wagon as now displayed at the Milestones Museum and showing the tipping bodywork.

Wallis & Steevens (Wallis & Steevens Ltd, Basingstoke, Hampshire)

Another company represented by a single surviving wagon is Wallis & Steevens. No. 7279 (AA 2470) dates from 1912 and is also part of the Milestones Museum collection. It was displayed at the Great Dorset Steam Fair in 1994 when there was a wagons theme. This was new to Pickford's as fleet No. 104 and had five more owners before preservation beckoned.

4
Undertype Wagons

Atkinson (Atkinson & Co., Preston, Lancashire)

Atkinson developed from a wagon repair company and agents for Alley & McLellan (Sentinel) to building their own wagons from 1916 onwards. A total of 325 were built until production ended in 1929 in favour of trailers and then diesel lorries from 1931 onwards.

A wagon that has been with the founder of Preston Services, Mr List-Brain, for many years is Atkinson 6-ton Colonial type No. 72 of 1918. The remaining parts were imported from Australia in 1976 and restored by Tom Varley. A three-way tipper, it is the only known survivor of the wagons built by Atkinson & Co. This was seen at the rally at Preston, near Wingham, Kent, in 2009. Although the wagon was used in the Australian goldfields area, the present livery is fictitious and was applied when the wagon was restored.

Foden (Fodens Ltd, Sandbach, Cheshire)

For years Foden had persisted with overtype wagons with their in-line boilers and cylinders atop, traction engine style. This all took up valuable load space. They developed the Speed 6 and Speed 12 undertype wagons with vertical boiler, as built by rivals Sentinel in a last-ditch attempt to promote the steam wagon. But it was too late, the diesel lorry had arrived and legislation discriminated against steam. After 1933 Foden would switch entirely to making diesel lorries and buses.

2016 was notable for the completion of not just one but both of the surviving Foden undertype wagons. These both were entered at the Great Dorset Steam Fair. LG 4815 is a Speed 6 model of 1930, which was bought totally derelict from the sale by W. J. King, contractors and quarry owners of Bishops Lydeard, Somerset, in May 1988 – a source of many Foden and Sentinel wagons.

The Speed 12 wagon MI 3304 is No. 13976 of 1931, which originally worked in Ireland. The Speed 6 and Speed 12 engines had a totally enclosed duplex engine, two-speed gearbox, shaft drive and the boiler worked at a pressure of 275 psi. The official top speed was 45 mph. These were advertised with pneumatic tyres from the outset.

Fowler (John Fowler & Co. (Leeds) Ltd)

Although Fowler were prolific builders of all types of traction engine, with ploughing engines being a speciality, wagons were a mere sideline and only one example survives in the UK. Production only started in 1924 when petrol lorries were already well established, with a total of 124 being built until 1931.

No. 19708 dates from 1931 and was built originally as a gully emptier for the City of Leeds. Withdrawn in 1941, it was rescued by Tom Varley in 1968 as a pile of cut-up parts and fully restored. This was it at Peterborough in 1982.

Fowler No. 19708 again, in a later livery at the 1996 GDSF. This is fitted with a V-twin compound engine.

Garrett (Richard Garrett & Sons Ltd, Leiston, Suffolk)

Garrett made both overtype and undertype wagons until 1931. This is No. 34841 of 1926. In part-restored form, the position of the boiler can readily be seen. Note the lagging to protect the crew from some of the heat from the boiler. Harewood House, Yorkshire, 1997.

No. 34841 is a QL class 6-ton wagon and is fitted with duplex cylinders and piston valve gear. Ackerman steering is fitted and the final drive is by chains to the rear axle. Weeting, 2001.

In this later view from 2012 at Old Warden the body has been rebuilt as a dropside design, while the wheels have received disc covers.

The asymmetrical cab on CV 5166, No. 35465 of 1931, is a distinguishing feature. Another QL class wagon, this was supplied new to Glover & Uglow Ltd, Callington, Cornwall, later passing to Cornwall County Council and ending its days in a Devon quarry. It is fitted with a tipping body. Alexandra Palace, London, in 1980.

Leyland (Leyland Motors Ltd, Leyland, Lancashire)

This company was founded as the Lancashire Steam Motor Company in 1896. The first steam wagons had vertical boilers fired by petrol but thereafter coke was used. The class H 5-ton wagon was produced from 1904 to 1911 when it was joined by the class F, which had a superheated boiler and poppet valves rather than slide valves. In 1905 the company started to build petrol internal combustion engine vehicles and steam wagon production ceased in 1926, with remaining spares being transferred to Atkinson's at Preston.

At Peterborough in 1981 is this unregistered 1923 Leyland F2 wagon. This was found derelict in Australia, shipped home in 1968 and restored over a two-year period by apprentices at Leyland Motors. This is now housed in the British Commercial Vehicle Museum at Leyland.

Sentinel (The Sentinel Waggon Works Ltd, Shrewsbury, Shropshire)

This company of Alley & MacLellan Ltd, Glasgow, first produced steam wagons (or 'waggons' as they described them) in 1906. In 1917 they were renamed the Sentinel Waggon Works Ltd and moved to Shrewsbury. They would become the most successful maker of undertype wagons, selling both to home and export markets. A few overtypes were also made but only from 1911 to 1912. Steam wagon production continued right up until 1951 (for export), with diesel lorries made from 1947 onwards.

V 3057 is No. 753 of 1914, and thus built at Glasgow. This is the second oldest survivor in the UK and is now kept at the Grampian Transport Museum at Alford. However, I encountered it at the 1998 Great Dorset Steam Fair. Note that this has a cab roof and back wall – on the oldest example No. 190 of 1907 the driver is completely exposed to the elements, like on the Hindley wagon.

V 513 (No. 1170) *The Old Man* was rebuilt using the chassis of No. 1170, dating back to 1907, and parts of No. 3748. The chassis was also extended to 8-ton size. Taken at Enfield in 1990.

AW 3407 is No. 1488 of 1917. This and the following examples were all built at Shrewsbury, and this carries a Shropshire registration number. When seen at the Great Dorset Steam Fair in 1990 it carried this dray body and was liveried for Bass & Co. brewery. Its last commercial user was Brown Bayley Steelworks where it was No. 8 in their fleet.

AW 3407 seen again at Old Warden in 2014. It was now named *Denby Flyer* and fitted with a dropside body.

Also dating from 1917 is No. 1716 *Unrepeatable* (AW 3833). This has also had its bodywork rebuilt in preservation since photographed here at Marks Tey, Essex, in 1989. I later saw it in 1995 without the roof cover and in the colours of Charringtons Coal & Coke Merchants.

AW 3835, also of 1917. All these are Standard models which featured a water tube boiler working at a pressure of 230 psi, a two-cylinder engine and a single chain drive. A total of 4,317 Standard wagons were manufactured up to 1923. This is part of the Science Museum reserve collection housed at Wroughton near Swindon, where it was taken in 1985.

Weeting in 2008 and Sentinel Standard wagon AW 4131 of 1918 takes on water from a hydrant in Fengate Drove, the road that leads down towards nearby Brandon. The tank on the wagon would be used to replenish engines on the rally field.

AW 8451, No. 3549 of 1920. Only the chassis had survived, in use as a trailer, but the wagon was restored over fifteen years using parts from other engines. Old Warden, 2007.

FA 1803/KC 5071, a matching pair. Both these engines were built in 1924. Nos 5407 and 5256 are examples of the Super type, introduced in 1923, which featured a double chain drive to the rear axle. They also had a new differential gearing, redesigned steering, enclosed cabs and a new boiler design. A working pressure of 230 psi could be raised in forty-five minutes from lighting up. These engines carry the livery of the United Africa Co. Ltd of Liverpool, one of the last companies to work these wagons commercially until the 1960s. They were posed together at Old Warden in 1998.

Displayed in the colours of its original owners from the Isle of Wight, DL 5478 (No. 6887) dates from 1927 and is fitted with a tipping body being demonstrated at Old Warden in 1998. It was taken out of service in the 1930s due to the prohibitive road tax and sent to a scrapyard at Cowes from where it was rescued in the 1960s.

YC 7914 (No. 8109) was built in 1929 and is seen at Rushmoor Arena in 1988. It was originally built for use by Henry Butt & Co. in a Mendip quarry as a tipper wagon where it spent all its working life, never being fitted with pneumatic tyres.

Similar is No. 7591 (WE 3236) of 1928. This was new to George Senior, Sheffield. It ended its life at the nearby Brown Bayley Steel Works where it was the only Super Sentinel alongside the earlier Standard models. Comparing these nearside and offside views shows the separate chain drives on either side of the chassis frame. Blackpool Airport, 1991.

KA 5574. While the previous two wagons were 5-ton models, No. 6725 (KA 5574) is a 6-ton version built in 1927. This was seen in the sun at Old Warden in 2003, carrying the livery of its original owners.

YI 8056. Another 6-ton Super wagon is No. 6079 of 1925. Originally registered in Ireland, this is now preserved in Hampshire and was an entrant at the Netley Marsh Rally near Southampton in 2022. Sentinel first offered pneumatic tyres in 1931, so this would have been built originally with solid tyres.

Also a 6-ton model is NE 4770. When seen at Little Wymondley, Hertfordshire, in 1989 it was carrying the livery of sugar refiners Tate & Lyle and with a realistic sheeted load. Although this wagon never ran for Tate & Lyle Ltd, they did have two similar wagons. NE 4770 was new to an operator in Manchester.

Sentinel Super wagon No. 7651 (now BS 9215) was supplied in 1928 to Cambridgeshire University and Town Gas where it worked until the 1950s before being scrapped. The wagon as we see it today was rebuilt from remaining parts over a seven-year period and is seen posed at Old Warden with other Sentinels.

Most recent of the Super Sentinels to survive is No. 8393 (DX 9048) of 1930. This was new to W. M. Brown for transporting flour. In 1933 it went back to the Sentinel factory to be fitted with pneumatic tyres and electric lighting. In 1957 it was purchased by Wingham Engineering and restored in preparation for the Suez crisis but never used. It was loaned to HMS Sultan, the Royal Navy School of Marine and Air Engineering, Gosport, in 1960 and purchased by them in 1970. It is maintained and rallied by a volunteer crew of serving and ex-service personnel and civilians. Rushmoor Arena, 1990.

AW 3321. Although this is clearly badged as a Sentinel Super, AW 3321 is a bit of an anomaly. In the Traction Engine Register it is listed as a 'Standard' model and carries maker's number 1465 in the Standard series from 1916. It was returned to Sentinel's in the late 1930s and was rebuilt in 1942 as a 'Lightweight Super'. It was then used as a works transport vehicle until 1948 when it was sold to Scientific Roads and became a tar sprayer before entering preservation in 1957. The Lightweight Super combined the best features of the Super and DG models to compete with petrol lorries providing 100 hp at a cruising speed of 40 mph. Seen participating in the parade at the 1990 Knowl Hill Rally.

NT 8347. Taken in Battersea Park at the Wheels of Yesterday Rally in 1984, this 1926 Sentinel Super tractor and trailer had been on the HCVS London–Brighton Run the previous day. Restoration was completed between 1967 and 1971. The Wheels of Yesterday Rally began in 1980 and would switch to Crystal Palace when the departure point for the Brighton Run moved there in 1989 but it declined and ceased in 1995.

PD 1854 is a Super Sentinel dating from 1924. It started as a wagon with Hodgsons Brewery, Kingston-upon-Thames. Ten years later it passed to an owner from Liverpool and was converted to a tractor by the Sentinel service garage in Liverpool. It remained in commercial use with Criddle & Co. until 1962 before being bought for preservation. This view is at Shrewsbury in 1985.

Although looking like a typical Super tractor, SV 5525 only dates from 1996. It was made up then using the boiler from Sentinel railway loco No. 9593 and a new chassis/body. It carries the maker's plate for No. 5335. This was taken at Weeting in 2001.

Sentinel introduced a single-speed timber tractor in 1923, which used the boiler from the Super Sentinel wagon and a direct chain drive to both rear wheels. A two-speed model followed in 1925. The second of these was No. 6426, which was exported to Australia in 1926. This was repatriated from Australia in 2006 and appeared at Old Warden in 2010. It has since been exported to America in 2017.

The rear end of No. 6426 (BF 5417) showing the powerful winch.

From 1927 to 1932 the Super series was replaced by the DG series, a double-geared design with Duplex cylinders. These were in production until replaced by the final S series shaft drive wagons. No. 8595 appeared at the Commercial Motor Show in 1931. It was then bought by Henry Franklin Ltd, millers from Biggleswade, Bedfordshire. It was used for flour deliveries to London and collection of wheat from the docks, working with a Sentinel trailer to give a total carrying capacity of 16 tons. During the war longer journeys were made to such places as Liverpool and Cardiff. When withdrawn in 1953 it had covered over 700,000 miles. It was preserved in Canada from 1958, returning in 2004. Here it is at Old Warden in 2016, close to its old working area.

KG 1123 is No. 8714 of 1932. When photographed at Crich in 1983 it carried this flatbed body and the livery of its original owners. At an earlier time in its history this wagon had moved to Ireland and had been converted to burn peat.

In 2003–04 KG 1123 received this replica omnibus body based on an original vehicle constructed in 1924 and was used on tours in the Lake District. In 2012 it was purchased jointly by the Saunders family from Stotfold, Bedfordshire, for their collection and the Bressingham Steam Museum for carrying passengers during the tourist season. This is seen at Stotfold Mill in 2014.

6-ton DG4 No. 8084 was built in 1929 for Hovis Ltd, working out of Trafford Park, Manchester. This was one of fourteen Sentinels operated by Hovis including two DG4s but is the only survivor. In 1933 it was sold to Grieve Haulage Co. Ltd of Bromsgrove, who had it converted from solid to pneumatic tyres. This also forms part of the Saunders collection. The remaining parts were bought in 2003, work began in 2007 and restoration was completed in 2012. The current boiler is believed to have originally been fitted as a replacement in an LNER Sentinel shunting locomotive.

1931 Sentinel DG4P type No. 8571 (KF 6482). No. 8571 was supplied new to Samuel Banner & Co., Bootle, passing to Paul Bros, Birkenhead, in 1936, with whom it would remain in service until 1949 when it passed into preservation with Edgar Shone. In 1977 it was sold on and painted in the colours of Morris, Shrewsbury, who once used similar wagons. It was looking quite at home against the background of railway arches while taking part in the May 1984 Birmingham Museum of Science & Industry gathering road run around the back streets.

A number of wagons finished their working lives as tar-spraying vehicles, but OD 1572 is a rare example of one preserved in this form. This DG4P dates from 1932. Now preserved in Sussex, this was at a rally at Lingfield in 1997.

DG4 wagon TE 9662 dates from 1929. It was new as a three-way tipper to Hugh Winterbottom, Coal Merchants, Oldham. Next passing to Heaton Mills, Middleton, it ended commercial life with Glossops, a tar road contractor. It appeared at St Albans in 2015 following a four-year restoration.

The lettering on 1929 Sentinel DG6 wagon No. 7966 is a credit to the art of the signwriter. Only the chassis and body had survived when acquired by Barry Cousins in 2001. Restoration was completed in 2009 and it was displayed at Old Warden in 2010. This was the last of thirty-two Sentinels supplied to J. Lyons & Co. and the only DG6 model. It only worked until 1933, a victim of prohibitive speed legislation on solid-tyred vehicles.

The most recent DG series engine in the Register is DG6 No. 8803 (YD 7012) built in 1933. This was another of the wagons that began service with W. J. King, Bishops Lydeard, and was ordered with solid tyres rather than pneumatics. It has been with several different owners in preservation and has been restored to the original livery.

RG 1417 is a 1930 Sentinel DG6T wagon with a three-way tipping body supplied new to William Towse Ltd of Aberdeen. It was sold to Aberdeen Harbour Board in 1937 to carry concrete blocks used to repair the breakwater. First sold into preservation in 1965, it had been extensively restored by the present owner during the previous five years. RG 1417 is depicted at Brighton in 2006.

1930-built DG6 GF 8655 taken at Quainton Road station, home of the Buckinghamshire Railway Centre, in 1998. Although now fitted with a dropside body, this was built as Britain's first mobile concrete mixer for British Steel Piling Co., London. It later passed to Willments, Twickenham, and was used as a stationary mixer before passing into preservation. It was initially rallied in its mixer guise before later being rebodied as seen.

No. 8590 (GT 2827) *Elizabeth* started as a flatbed wagon with the Cement Marketing Co., before ending its days as a tar sprayer. It was fitted with this detachable bus body in preservation and spent some years at Whitby in the ownership of Vern and Viv Smith where it operated passenger trips around the town for much of the year. In November 2010 it travelled to London and took part in the Lord Mayor's Parade, promoting Yorkshire tourism (the new Lord Mayor was a Yorkshireman). It also returned in 2011. It later passed to new owners at Weston-super-Mare in 2014, who rebuilt it with a new detachable bus body and ran it on tours there.

Sentinel also sold eight TDG timber tractors to the British market. These were fitted with a separate steam winch engine for dragging and loading felled trees. Four of these survive. Three of the survivors worked for R. M. Woolley of Bucknall, Shropshire, into the 1960s. No. 9097 (ARE 1950 was new to A Wooton & Sons, Cannock, and is now part of the Claude Jessett Collection from Hadlow Down near Uckfield, Sussex. This was working in the timber sawing demonstration area at the 1996 GDSF.

Old Bill, JB 1655 of 1933, was owned by R. M. Woolley from 1947 to 1964. This was taken in the ring at Rushmoor Arena, 1986.

Another of the R. M. Woolley tractors was No. 8756 (VN 4294) *Brutus* of 1933, which became part of the Bressingham Steam Museum collection. It visited the Weeting Rally in 2003. The first commercial owner was T. Place & Sons Ltd, Northallerton.

When photographed again at Weeting in 2007 the livery had changed as shown to that from its final working life. This has since become part of the collection of Michael List-Brain at Preston, Kent, and has been painted green.

The Fourth TDG survivor is No. 9236 (AAM 483). This was new to the Yonder Hill Saw Mills, Chard, but later returned to the Sentinel Works and the timber winch was removed. It was used as a works shunter during the Second World War and later in Southampton Docks. After passing through several owners, since 1995 it has been restored as a ballast tractor in the colours of Bishop & Sons and is seen with a matching trailer at Brighton in 1998.

At Peterborough in 1980, present from the Tom Varley collection, was FD 6603 *Pendle King*, a 1931 Sentinel DG6 wagon. Contrast its appearance with that of the same wagon when in its working days (see p. 7). It was supplied new on solid tyres to Dudley & Blowers Green Transport Co. but was shortly afterwards converted to pneumatics. It worked on long-haul trips to Manchester, London, etc., before, like many others, being sold on and ending commercial life as a tar sprayer.

There are only two eight-wheel Sentinels listed: this DG8P, No. 8016, and one S8 type (see p. 78). UX 5355 dates from 1929 and was rebuilt from surviving parts but requiring a new chassis. It was new to Tarmac as a solid-tyred six-wheel tipper but returned to the Sentinel works a year later for conversion to eight wheels and pneumatic tyres, then being used until 1945. It is posed for photographs at Old Warden in 2018.

UX 8724 is a unique wagon, being the prototype shaft drive engine built in 1930 and classed as type SDDG4. It has steam brakes operating on all wheels. This worked for Grieve Haulage Co. of Birkenhead from 1931 to 1947. This view was at the Chiltern Steam Rally, Prestwood, in 2014.

The production shaft drive wagons from 1933 became the S type and featured a new cab design with the boiler behind the crew. HV 3865, seen at Bloxham in 1992, is a wagon that originally hailed from my local area of east London. The HV registration is from the County Borough of East Ham where E. & A. Shadrack Ltd were coal merchants. Sentinel S4 No. 9016 dates from 1934.

S4 wagons 9031 and 9032 of 1934 carry consecutive registration numbers and both worked for the Cowlairs Co-operative Society. They were displayed together at Old Warden in 2009. No. 9031 (US 5431) was preserved in the USA between 1961 and 2008. No. 9032 has completed the journey from John O'Groats to Land's End and also steamed all around Holland. Therefore, travelling from its home in Bath to Bedfordshire under its own power was no problem.

Seen at Brighton in 1987, this 1933 Sentinel S4 steam wagon, No. 8827, was entered by its preservationists in Holland. It has since been repatriated and regained its original UK registration UJ 2225.

Sentinel S4 wagon No. 8827 was in Holland from 1981 to 1998. Now back in the UK, it had been repainted into the colours of the Gas Light & Coke Co. when it featured at the Essex Steam & Country Show at Barleylands, Billericay, in 2010. A fully loaded S type could travel about 57 miles between water stops and about 190 miles on the 5 hundredweight of coal carried. Overall, their performance was superior to contemporary petrol and diesel lorries.

S4 wagon No. 9276 was new in 1936 to the Wingham Engineering Co. of Kent as fleet No. 72. It was used on brewery haulage for Whitbreads until replaced by a diesel lorry in the 1950s. It was sold for scrap to Hardwicks of Ewell and then passed to John Keeley in 1956 for preservation where it featured at the Knowl Hill rallies. After his death it was purchased by the Davenport family and returned to rallying after overhaul in 2022. Now with Rory Holbrook and fully repainted in his colours, it is seen at the 2023 Weeting Rally.

No. 8843 (UJ 2112) of 1933 was built as a three-way tipper. When purchased for preservation in 1977 it had been stripped to rebuild another wagon and was missing all the tipping gear. Instead, it was restored as a timber tractor with a crane and winch, in which form it was recorded at the former Great Dorset site at Stourpaine in 1988. It has since been rebodied as a tipping wagon as built.

No. 8933 (WV 4705) was built in 1934 and supplied new to Mitchell & Sons, Downton, Wiltshire. It passed to Ely Ales of Cardiff in 1939 and remained in their ownership until 1954 when it was purchased privately. It passed to new owners in 1991, who restored it over the following eleven years. It was entered in the HCVS London to Brighton Run in 2003.

S4 No. 9151 (JH 9994) was fitted with this freelance bus body in the 1960s. Based in County Durham, it was recorded at Chester-le-Street in 1995. The body was based on that fitted to Super Sentinel No. 5102 in 1924, which failed to find a buyer and ended up being used by the Sentinel Silver Band. JH 9994 was previously a wagon supplied to Simpsons Brewery, Baldock, Hertfordshire.

Representing the six-wheel S6 type is FH 8870 of 1934. This was purchased by the present owner in 1986 and after renovation made its rally debut here at Tallington in 1997.

In 1978 Foden were the sponsors of the HCVS London–Brighton Run, but this entrant was from their main rivals Sentinel. BRF 200 is a 1933-built S6 model with a tipper body used for carrying roadstone for Tarmac. It later passed to Cambridge Gas Works and was last used commercially during the 1956 Suez crisis when there was a shortage of oil fuel.

Another S6 is No. 8928 (JG 4222) of 1934. This was new to W. Hooker of Canterbury. It has been with the present owner since 2000 and was photographed at Old Warden in 2003.

1934 Sentinel S8 wagon No. 9105 (UJ 3652). This was built as a demonstrator and later worked for an operator from Newport, Monmouthshire, carrying steel plate from South Wales to Ford's factory at Dagenham. Seen heading for home and about to cross Chelsea Bridge after being an entrant at the Wheels of Yesteryear Rally in Battersea Park on 5 May 1986.

Since 2000 the S8 has been with R. Hazell & Sons along with S6 wagon JG 4222 and S4 wagon BEV 466. This was also taken at Old Warden in 2003. Only nine S8 wagons were built, so this is a lucky survivor.

Thornycroft (John I. Thornycroft & Co. Ltd, London)

John I. Thornycroft, a shipbuilder of Chiswick, London, built his 'Number One steam van' in 1896 as Britain's first steam wagon and this was exhibited at Crystal Palace in 1896. It featured a water-tube boiler and a twin tandem engine both of marine practice. Drive was by double chains to the front wheels while steering was achieved by the rear wheels, features discontinued on the subsequent wagons built. The load capacity was 1 ton, and maximum speed was 12 mph.

The Thornycroft first steam van was retained by the company and passed with the makers into the British Leyland Motor Corporation. After spending some years at the National Motor Museum at Beaulieu it now resides at the British Commercial Vehicle Museum, Leyland, where it was seen in 1989.

From the Robert Crawford stable came this 1900-built Thornycroft wagon, maker's number 39. This was originally owned and restored by the late John Crawley. It was seen at the Expo Steam event at Peterborough in 1980, which had a wagons theme.

The third survivor is No. 115 of 1902, also seen at Peterborough, although this time in 1981. This is now displayed at the Milestones Museum at Basingstoke, where Thornycroft & Co. Ltd later relocated. It was previously located at the British Commercial Vehicle Museum.

5
Transverse Boiler Wagons

Yorkshire (Yorkshire Patent Steam Wagon Co. Ltd, Hunslet, Leeds)

Yorkshire were the third largest maker after Foden and Sentinel, producing between 1901 and 1937. Oldest of the surviving Yorkshire wagons is No. 117, built in 1905. This is a 2-ton model with compound cylinders. Taken at Peterborough in 1982 when restored by Tom Varley.

No. 652 dates from 1914 and is a 6-ton WE model fitted with a box van body.

Dating from 1917, No. 940 *Pendle Queen* is a 3-ton WA class wagon with a flatbed body. This was supplied new to Clayton, Son & Co. of Leeds. When rescued in 1961 it had been laid aside since 1931, cut into pieces and thought to be unrestorable. This was also taken at Peterborough but in 1981.

Dating from 1927, No. 2118 is a WG model tractor, originally with an articulated trailer for the Leeds Electricity Department, and later rebuilt as a ballast tractor. Facilities for the crew have advanced with an enclosed cab now provided.

When photographed again in 1999 No. 2118 had been modified in preservation to have pneumatic tyres, giving a more modern appearance, although detracting from originality.

Also from 1927 and fitted with pneumatic tyres is No. 2108. Like Nos 117, 940 and 2118, this was part of the collection of Tom Varley from Gisburn at the time this photo was taken in 1981.

No. 2108 later changed hands and is now based in Somerset. This is at the Great Dorset Steam Fair in 1995.

6
Home-made Freelance Wagons

As well as the genuine wagons that have survived, there have been a number of freelance designs that have been created in recent years, perhaps utilising the chassis from a motor vehicle.

An interesting entrant at the Birmingham Museum of Science & Industry, Newhall Street in 1984 was POA 987, a home-made steam delivery van. This was created in 1973 using a 1905 van body and a 1901 Locomobile steam car engine coupled with a vertical boiler.

This one-off steam cart was built in 1895 by a Scottish postman named Mr Lawson. It has a single cylinder and a wooden frame. This is normally housed in the Grampian Transport Museum in Scotland but on this occasion in 1989 was visiting a 'Museums on the Move' event at Duxford Airfield.

Typhoo was constructed in 1970 and is based on an ERF lorry chassis fitted with a boiler from a Sentinel rail shunting engine. This was visiting the Gloucestershire Steam & Vintage Extravaganza in 2016.

Known as *Figment of the Imagination*, this is a freelance steam bus. This view at the Great Dorset Steam Fair in 1996 shows it when the bodywork was not yet complete so the framing can be seen. A vertical boiler is mounted behind the driving position.

This shows *Figment of the Imagination* when the bodywork had been completed. Note the nearside entrance with step access. It has a 1970s Leyland lorry chassis and a boiler from a 1900s steam launch. Taken at Old Warden in 2006.

This freelance design owned by Mr G. Needham is based on a 1940s Bedford ML chassis. This is fitted with a 6-inch scale McLaren boiler. The engine is a Daihatsu two-cylinder engine, chain driven through a four-speed Bedford gearbox. This was also taken at Old Warden, but in 2016.

It's called *Bitsa*, and that is what it is. This is a freelance showman's steam waggon owned by Robin & Rachel Neave of Alderholt. Originally built in the 1970s and acquired by the present owners in 2009, it has been rebuilt and now features a vertical boiler and a Bryan Donkin engine, as well as several Foden parts. Taken at Netley Marsh in 2014. Behind it comes Foden tractor KX 3340 *Samantha* of 1929.

I believe the chassis of this freelance pick-up started out as an Austin FX3 taxi. It was made in the 1950s but was later with an owner from Hampshire. It was spotted at Rushmoor Arena, near Aldershot, in 1991.

Known as the Spider tractor, KRO 284Y was built by a Mr Kilgour at the Silsoe Agricultural College, Bedfordshire, in 1984. The aim was for use by farmers in Malawi. It has a Lancashire type boiler, single cylinder and two speed transmission. This was taken at the Bedfordshire Steam Rally at Roxton Park in 1991.

A second, larger vehicle built by Mr Kilgour is Q885 NVS. I believe the chassis for this was an ex-army Bedford lorry. A two-cylinder high-pressure engine has a chain drive to a four-speed gearbox. The boiler can burn either coal or wood. Maximum speed was around 25 mph. This was at Old Warden in 2002.

This contraption, an entrant at Prestwood in 2023, is known as *The Dumper*. Built in 2001, it comprises a dumper chassis, an engine recovered from a factory in Luton and a 1966 boiler built by W. Gower in Bedford. The canopy came off an old milk float.

Seen at Redhill in 1992, this looks as though it has been built on an old electric milk float chassis. The vertical boiler is behind the driving cab.

This home-made wagon was seen going round the ring at Brockwell Park, London, in 1977.

This delightful contraption with its matching trailer carriage was seen at Horstead Keynes in 1993. I can imagine this trundling up and down a seaside promenade giving rides. Although not clear in this view, the wheels are of the wire spoked motorcycle variety.

I presume this started life as a 1930s light car, before being rebuilt as this steam-powered pick-up. It was on show at the 2009 Weeting Rally.

7
Miniatures

If you can't afford the real thing or haven't enough space, there is always the option of a working miniature. These can vary from around quarter size to three-quarters size. They may be completely scratch-built or assembled from a kit of parts. Some may represent types that have not survived in full-size.

Passing the line-up of engines at Shefford in 1994 is one of a series of six 6-inch scale miniature Foden wagons built by the Maskell family in their workshops from 1992 to 2000. L59 GNM was built in 1993 and is a replica of the Foden Brass Band Bus.

The theme at the Bedfordshire Steam & Country Fayre for 2003 was 6-inch scale miniatures and here some of the Maskell Foden wagons were posed up in front of the house for photography. There were over thirty half-size exhibits, four of which were shown with their full-size counterpart.

A superb miniature of a Foden showman's box van, and some excellent signwriting too! Seen at Singleton, Sussex, in 1999. Some seven of the surviving Foden wagons saw later use by showmen, although probably not with a dynamo fitted, although Garrett BJ 1366 was so fitted at one time (see p. 38).

The proud owner drives his miniature Foden wagon and trailer around the grounds of the Royal Gunpowder Mills, Waltham Abbey, in 2008.

This is a 6-inch scale version of a 5-ton Garrett overtype superheater fitted wagon, which was built by owner A. Putterill from original works drawings scaled down. This was at Old Warden in 2014.

A Mann wagon, probably three-quarter scale, at the GDSF in 2015.

This is a three-quarter scale miniature of an undertype Mann wagon. This was taken at a rally at Marks Tey, Essex, in August 1989. The vertical boiler would be ideal for brewing that cuppa, as well as powering the wagon!

The same wagon after probably changing ownership and with the bodywork modified somewhat. This was taken at Chatham Dockyard in 2003.

Seen at Redhill in 1993, this claims to be a Beyer Peacock design. It is probably built to three-quarters full size and has a transverse-mounted boiler.

Also at Redhill on the same day, this is a miniature of a Fowler wagon, of which only one full-size example survives.

At Old Warden in 2009 was this miniature of an early type of wagon. I am not sure what make is being represented but as can be seen the facilities for the crew would have been minimal. Also note the wooden brake block acting on the rear wheel and two chocks hanging from the body to act as a parking brake.

Acknowledgements, Bibliography and Further Reading

Thanks to Online Transport Archive for permission to use photos from the John Meredith Collection.

BATTEN, Malcolm	*Traction Engine Rallies: An Appreciation Over Seventy Years 1950–2019* (Barnsley: Pen & Sword, 2023)
COLBECK, Simon (ed.)	*Old Glory Archive Vol 5: The Colour Files* (Yalding: Kelsey Publishing, 2021)
JOHNSON, Brian	*The Traction Engine Register, 13th Edition* (Horsham: Southern Counties Historic Vehicles Preservation Trust, 2020, and earlier editions)
JOHNSON, Brian	*Steam Traction Engines, Wagons and Rollers in Colour* (Poole: Blandford Press, 1976)
LOCKETT, David & ARLETT, Mike	*Traction Engines: A Colour Portfolio* (Hersham: Ian Allan, 2002)
SAWFORD, Eric	*Steam Power, Farm & Highway* (Burton-upon-Trent: Trent Valley, 1988)
SAWFORD, Eric	*Steam Wagons in Colour* (Shepperton: Ian Allan, 1997)

Old Glory (Yalding: Kelsey Media), monthly magazine published since 1988

Steaming (Redditch: NTET), quarterly magazine of the National Traction Engine Trust

Vintage Spirit (Cranleigh: Steam Heritage Publishing), monthly magazine published since August 2002

Rally programmes from events I have visited.